Complete Guide on April 2024 Total Solar Eclipse

How to watch, understand and stay safe on April 8

Pamela K. Brown

Before this document is duplicated or reproduced in any manner, the publisher's consent must be gained. Therefore, the contents within can neither be stored electronically, transferred, nor kept in a database. Neither in Part nor full can the document be copied, scanned, faxed, or retained without approval from the publisher or creator.

Copyright © by Pamela K. Brown 2024. All rights reserved.

Table of Contents

Introduction   4
   The Phenomenon of Solar Eclipse   4
   What Makes the April 2024 Eclipse Special?   8

Chapter 1: Understanding Solar Eclipses   13
   The science of eclipses   13
   Types of Solar Eclipse   18
   The Path of Totality   23

Chapter 2: Historical Eclipses   29
   Significant Eclipses Throughout History   29
   Myths and Legends about Eclipses   33

Chapter 3: Preparing for the Eclipse   37
   Choosing the Best Location   37
   Equipment Checklist.   42
   Plan Your Eclipse Trip   47

Chapter 4: Safety First   53
   How to safely see a solar eclipse   53
   Protecting your eyes   57
   Common Misconceptions and Dangers   62

Chapter 5: Photography and the Eclipse 67

Photographing the Eclipse - Tips and Techniques 67

Required Equipment for Eclipse Photography 72

Chapter 6: The Eclipse Experience 77

What to expect during the eclipse 77

Personal Accounts for Totality 81

The Emotional Effects of Eclipses 86

Introduction

The Phenomenon of Solar Eclipse

A solar eclipse is one of the most fascinating and significant natural phenomena. As an enthusiastic skywatcher, I have always found this event to be a humbling reminder of our place in the cosmos. To me, a solar eclipse is more than simply a rare astronomical event; it's a celestial dance, a cosmic alignment of the Sun, Moon, and Earth that illuminates the inner workings of our solar system.

Solar eclipses have captured people's imaginations from time immemorial. These occurrences were long cloaked in myth and mystery, with many people interpreting them as omens or heavenly interventions. Today, we see them as stunning coincidences of orbital mechanics, in which the Moon passes directly between the Earth and the Sun, producing a shadow on our planet's surface.

To properly appreciate the beauty of a solar eclipse, one must understand the physics underlying it. The Sun, a massive fusion reactor, is around 400 times bigger than the Moon. However, by a stroke of cosmic luck, it is around 400 times further distant from us, making the two bodies look nearly the same size in our sky. This size-distance ratio allows the Moon to cover the Sun so accurately, resulting in breathtaking moments of totality during a complete solar eclipse.

As a total solar eclipse occurs, the world around me changes. The sky progressively darkens, the temperature lowers, and nature becomes eerily silent. Birds stop singing, and animals act as if darkness had fallen. Then, at totality, the Sun's bright corona emerges, forming a halo of ghostly light around the veiled Sun. It's a sight that moves the soul, a short look into the Sun's outer atmosphere that is normally veiled from our eyes.

I've chased eclipses all over the world, and each time I felt a deep connection to the universe. Eclipse chasers share a sense of community and a love of the exquisite majesty of these natural phenomena. We assemble, wait, and marvel together, knowing that for a few precious minutes, we are one under the Moon's shadow.

The anticipation for the April 2024 complete solar eclipse is already mounting. This event promises to be a stunning display for those on the path of totality. I can't help but feel excited at the prospect of watching another ballet of shadows, another moment when the Sun, Moon, and Earth come together to produce a natural wonder.

As I prepare for the approaching eclipse, I think of the many people who have stared up in amazement at eclipses throughout history. Solar eclipses have inspired amazement and interest among astronomers, scientists, poets, and philosophers across nations and centuries.

In writing this, I intend to explain not just the mechanics of solar eclipses, but also the emotional impact they have. They are reminders of our transient life, invitations to halt and gaze up, and chances to connect with something bigger than ourselves. As the April 2024 eclipse approaches, I invite everyone to take in the awe of this celestial event by standing in the Moon's shadow and observing the Sun and Moon's spectacular dance.

For those who have never seen a solar eclipse, I recommend that you be ready to be moved. It's a phenomenon beyond words, a narrative spoken in the language of light and shadow. And for those little moments of totality, we are all a part of that tale, here on this small, blue planet, watching the cosmos unfold overhead.

What Makes the April 2024 Eclipse Special?

The April 2024 complete solar eclipse is more than simply a celestial event; it's a great sight that has captivated millions of people across the globe. This eclipse is unusual in terms of its course, duration of totality, and timing within the solar activity cycle, making it an especially exceptional occasion for both casual onlookers and scientists.

A Rare Path Through North America

The path of totality for the April 2024 eclipse will cross North America, from Mexico to the United States and Canada. This track is crucial because it will pass over heavily populated areas, allowing millions of people to see the totality without having to go far. Cities such as Dallas, Cleveland, and Buffalo will get the full impact of the eclipse, making it easily accessible.

Duration of Totality

The duration of totality during the April 2024 eclipse is one of its most notable features. The eclipse is predicted to last almost 4½ minutes at its peak, far longer than the Great American Eclipse in 2017. The longer time of darkness will allow spectators to fully appreciate the eclipse and scientists to study the Sun's corona[4].

Coinciding with the Solar Maximum

The period of the eclipse coincides with the solar maximum, the apex of the Sun's 11-year activity cycle. During this phase, the Sun's magnetic activity rises, making the corona more visible and increasing the possibility of detecting events like coronal mass ejections. These circumstances are likely to create a more dramatic and visually beautiful eclipse.

Cultural and emotional significance

Beyond their scientific relevance, solar eclipses have significant cultural and emotional meaning for many people. They are moments that transcend regular existence and provide a look into the magnificent grandeur of the cosmos. Because of its rarity and wide range of visibility, the April 2024 eclipse is expected to inspire intense emotional reactions and leave long-lasting memories for those who witness it.

A boon to scientific research

The eclipse provides a unique opportunity for scientific inquiry. Technological breakthroughs will provide researchers with more instruments than ever before to examine the Sun's atmosphere. Data from the eclipse will enhance understanding of solar phenomena and space weather, potentially impacting satellite communications and power networks on Earth.

Educational Opportunity

Eclipses are also effective instructional tools, instilling curiosity and enthusiasm in astronomy and science in individuals of all ages. The April 2024 eclipse is anticipated to excite a new generation of scientists and astronomers, as schools and educational institutions plan special programs and events to commemorate the occasion.

Preparation is Key

Given the magnitude of the event, plans are already in place to handle the flood of tourists to the path of totality and guarantee that everyone may safely experience the eclipse. Communities along the trail are arranging activities, and experts are highlighting the significance of wearing good eye protection to avoid harm from direct sunlight.

Chapter 1: Understanding Solar Eclipses

The science of eclipses

The study of eclipses is an intriguing subject that integrates astronomy, physics, and geometry to understand the celestial mechanics that underpin these spectacular phenomena. Understanding solar eclipses necessitates an investigation into the complex dance of celestial bodies and the shadows they throw.

Celestial mechanics and orbital dynamics

At the heart of eclipse science is the study of celestial mechanics, namely the Earth and Moon's orbits around the Sun. The Earth circles the Sun in an ellipse, as does the Moon. These orbits are not exactly lined; the Moon's orbit is around 5 degrees tilted toward the Earth's orbital plane, known as the ecliptic.

Solar eclipses happen when the new Moon departs between the Earth and the Sun, creating a shadow on the Earth. The arrangement and distances between the three bodies determine whether we see a partial, annular, or complete eclipse.

Umbra and Penumbra

The Moon's shadow has two components: the umbra and the penumbra. The umbra is the innermost and darkest section of the shadow that obscures the Sun, resulting in a total eclipse. The penumbra is the outside edge of the shadow where the Sun is only partially obscured, resulting in a partial eclipse.

Saro Cycle

Eclipse prediction is based on knowing the Saros cycle, which is an 18-year period in which eclipses reoccur with identical features. The cycle occurs when the Sun, Earth, and Moon align in almost identical relative geometry.

This cycle has been known since ancient times and is still a useful tool for astronomers.

The Importance of the Sun's Size and Distance

Solar eclipses rely heavily on the apparent size of the Sun and Moon in the sky. Despite being around 400 times larger than the Moon, the Sun is also about 400 times farther away from the Earth. This coincidence implies that the Sun and Moon seem to be almost the same size in the sky, allowing the Moon to obscure the Sun during a total solar eclipse.

The Path Of Totality

The path of totality is the short stretch of the Earth's surface where viewers may see a total solar eclipse. Its breadth is usually approximately 100 kilometers, however it can vary. Observers in this route witness the whole eclipse, including the spectacular darkening of the sky and the apparition of the Sun's corona.

Sun's Corona

The corona is the Sun's outer atmosphere, which is often obscured by the dazzling solar disk. During a total solar eclipse, the corona appears as a pearly white crown around the darkened Sun. Studying the corona during eclipses has revealed important information on solar wind, coronal mass ejections, and the overall structure of the Sun's atmosphere.

Eclipses and Scientific Discoveries

Solar eclipses have traditionally proven significant in scientific research. For example, during the 1919 total solar eclipse, studies of star locations near the obscured Sun proved Einstein's general relativity theory by revealing the bending of light due to gravity.

Safety when seeing eclipses

Observing solar eclipses takes extreme caution to avoid eye injury. The only safe method to stare directly at the uneclipsed or partially eclipsed Sun is using special-purpose solar filters like "eclipse glasses" or hand-held solar viewers. Indirect approaches, such as pinhole projectors, offer a safer option.

The physics of eclipses demonstrates the regularity and beauty of the natural world. It uses observational astronomy, geometry, and physics to explain how these exceptional phenomena happen. Solar eclipses not only give a spectacular view but also offer a unique chance for scientific investigation, improving our grasp of the world we inhabit. As we continue to study eclipses, we learn more about our solar system and its rules.

Types of Solar Eclipse

Solar eclipses are among the most spectacular celestial occurrences, attracting viewers with their beauty and uniqueness. They occur when the Moon passes between the Earth and the Sun, briefly blocking the Sun's brightness. There are four types of solar eclipses: partial, annular, total, and hybrid. Each kind offers a distinct experience that requires precise conditions to occur.

Partial Solar Eclipses

A partial solar eclipse occurs when the Moon only partially obscures the Sun as seen from Earth. This sort of eclipse occurs when the Earth, Moon, and Sun do not line exactly. The Sun seems to cast a dark shadow over only a portion of its surface. Partial eclipses are more common than complete or annular eclipses and may be seen from a larger portion of the Earth.

Annular solar eclipses

An annular solar eclipse happens when the Moon is too far away from Earth to totally obscure the Sun. This creates a ring-like image, with the Sun's outer edges remaining visible and producing the "ring of fire." Annular eclipses need a certain alignment in which the Moon's apparent size is less than the Sun's, which occurs as a result of the Moon's elliptical orbit around the Earth.

Total solar eclipses

Total solar eclipses are the most magnificent, with the Moon covering the Sun and transforming day into night for a brief duration. Observers along the line of totality will notice darkness, a dip in temperature, and the emergence of the Sun's corona, a crown of plasma that is normally obscured by the sun's dazzling brightness. Total eclipses are uncommon in any one area because the line of totality is so narrow, often around 100 kilometers wide.

Hybrid solar eclipses

The most unusual sort of solar eclipse is a hybrid. They are a blend of total and annular eclipses at different points along their courses. At some spots along the eclipse path, watchers will witness a total eclipse, while at others, it will be annular. This occurs when the curvature of the Earth's surface causes various sites to enter and exit the umbra and antumbra (the areas affected by an annular eclipse) at different moments throughout the eclipse.

The geometry of eclipses

The occurrence of each form of solar eclipse is determined by the geometry of the Sun, Moon, and Earth. The distances between these celestial bodies and their respective positions dictate the sort of eclipse we will witness. The Moon's shadow is divided into two parts: the dark umbra, which completely obscures the Sun, and the lighter penumbra, which only partially covers the Sun.

Scientific Importance

Each sort of solar eclipse presents excellent chances for scientific inquiry. Total solar eclipses, for example, enable astronomers to conduct in-depth studies of the Sun's corona. Annular eclipses can help us comprehend the Sun's exact proportions and behavior throughout various stages of the solar cycle.

Cultural Impact

Solar eclipses have had a profound influence on culture and history. They have been viewed as omens or divine messages, and they have influenced historical events. Today, they continue to inspire amazement and wonder, attracting large numbers of enthusiasts and visitors to the totality trails.

Safety in Viewing

Regardless of the type, it is critical to view solar eclipses safely. Looking directly at the Sun without sufficient protection might result in significant eye injury. Special eclipse glasses or solar viewers are required to shield the eyes from dangerous sun radiation.

Understanding the many forms of solar eclipses is necessary for enjoying these natural events. Each eclipse, whether partial, annular, complete, or hybrid, affords a unique view into the workings of our solar system and a moment of connection with the universe. As we await future eclipses, we prepare not just to watch an astronomical spectacle, but also to engage in a shared human experience that transcends time and space.

The Path of Totality

The name "Path of Totality" evokes feelings of awe and expectation among people who follow the Moon's shadow over the Earth. The short passage allows onlookers to witness the entire majesty of a total solar eclipse, a celestial phenomenon in which the Moon obscures the Sun, turning day into darkness and displaying the Sun's magnificent corona. Understanding the Path of Totality is more than simply knowing where to stand on the day of an eclipse; it's about comprehending the cosmic choreography that permits such a thing to happen.

Orbital Mechanics & Alignment

The Earth, Moon, and Sun align precisely to create the Path of Totality. This alignment is unusual because the Moon's orbit around the Earth is inclined by around 5 degrees from the Earth's orbit around the Sun. As a result, the Moon's shadow rarely passes over the Earth. A complete solar eclipse occurs when the Moon is at one of two places in its orbit, known as nodes when the Moon's orbit crosses the ecliptic plane (the Earth's orbit around the Sun).

The Moon's Shadow

During a complete solar eclipse, the Moon creates two types of shadows on Earth: umbra and penumbra. The umbra is the deepest and darkest section of the shadow, where the Sun is entirely covered. Observers within the umbra experience totality. The penumbra is the outside edge of the shadow where the Sun is only partially obscured, resulting in a partial eclipse.

The track of totality

The Path of Totality is usually roughly 100 miles broad and can extend for thousands of miles over the Earth's surface. Its actual size and shape during any given eclipse are determined by a variety of variables, including the Moon's distance from the Earth and the curvature of the Earth's surface. As the Earth rotates and the Moon revolves, this route moves around the world, leaving a trail that eclipse chasers anxiously record and trace.

The duration of totality
The duration of totality on the Path of Totality might vary greatly. It can last anywhere from a few seconds to over seven minutes, but the majority of total eclipses endure two to three minutes at any place. The duration is greatest at the place where the axis of the Moon's shadow cone is closest to the Earth's center.

The Experience of Totality
A total solar eclipse provides an experience unlike any other. The environment undergoes significant changes as the Moon covers the Sun. The temperature lowers, animals behave as if night has fallen, and stars appear in the sky. For those few seconds, the Sun's corona appears like a halo around the Moon, a sight that has captivated humanity for millennia.

April 2024 Eclipse

The complete solar eclipse on April 8, 2024, is particularly noteworthy since its Path of Totality will travel across heavily populated areas, making it visible to millions. The eclipse will pass from Mexico through the United States and into Canada. This eclipse is also projected to coincide with the solar maximum, resulting in a more active and potentially more visually appealing corona.

Scientific Importance

The Path of Totality provides scientists with a unique opportunity to study the Sun's atmosphere, particularly the corona, which is typically masked by the dazzling solar disk. Observations done during totality can lead to new insights into solar physics, such as solar flares and coronal mass ejections, which can have a substantial impact on Earth's space weather.

Safety and Preparedness

Those planning to watch the eclipse must be well-prepared. This involves being in the Path of Totality and wearing appropriate eye protection, such as eclipse glasses, to safely witness the spectacle. Communities along the path frequently arrange viewing parties and educational programs, making the eclipse a shared experience.

The Path of Totality is a tiny path leading to one of nature's most spectacular spectacles. It's a journey that inspires us to observe the amazing alignment of celestial bodies and consider our role in the cosmos. Whether you are an enthusiastic eclipse chaser, a casual spectator, or a scientist looking for data, the Path of Totality is an event that promises amazement, discovery, and a shared human experience that crosses countries and cultures. As we approach the 2024 eclipse, anticipation grows for what will be a spectacular occasion for everyone who stands in the Moon's short shadow.

Chapter 2: Historical Eclipses

Significant Eclipses Throughout History

Solar eclipses have historically been significant astronomical events that have piqued the interest and imagination of civilizations all over the world. These celestial events have not only been breathtaking, but they have also played crucial roles in the evolution of astronomy, formed cultural narratives, and even changed the course of history.

The earliest records

The first reported solar eclipse goes back to 3340 B.C., where spiral petroglyphs at the Loughcrew Cairn L Megalithic Monument in Ireland reflect observations of such an occurrence. Moving ahead to 1375 B.C., engravings on a clay tablet discovered in Ugarit, Syria, depict a total solar eclipse, with the city experiencing a near-total phase.

Predictions and omens

In ancient times, predicting solar eclipses was seen as a strong tool for astrologers and a noteworthy accomplishment for astronomers. As early as 2500 B.C., the Babylonians established ways to foresee eclipses based on the Saros cycle, which is around 18 years after which comparable eclipses occur. Failure to forecast an eclipse, however, might have serious implications, as the Chinese astrologers Hsi and Ho were supposedly killed for failing to foresee the solar eclipse in 2134 B.C..

Eclipses in Mythology and Culture

Eclipses have frequently been viewed through the lenses of mythology and cultural beliefs. Solar eclipses, for example, were considered ominous omens for ancient Chinese and Babylonian monarchs. During eclipses, the Babylonians would appoint surrogate monarchs to safeguard the genuine monarch from divine vengeance.

Scientific breakthroughs

Solar eclipse observations have allowed for significant scientific discoveries. The complete solar eclipse of May 29, 1919 was a watershed moment for Albert Einstein's theory of general relativity. Sir Arthur Eddington, a British scientist, witnessed light bending around the Sun during the eclipse, confirming Einstein's prediction. Another scientific milestone was French astronomer Jules Janssen's discovery of helium during a complete solar eclipse on August 18, 1868.

Historical events

Solar eclipses have also affected historical events. The eclipse of 585 B.C. notably interrupted a fight between the Lydians and the Medes, with the darkened skies taken as a message to end hostilities and negotiate peace. The Greek historian Herodotus recorded this incident, emphasizing the influence of cosmic phenomena on human affairs.

Modern observations

Solar eclipses have remained fascinating and popular in recent years. The Great American Eclipse of 2017 was a well-publicized cultural event, with millions of people across the United States viewing the totality. Such events not only give a spectacle but also chances for scientific research and public involvement in astronomy.

Solar eclipses have been crucial events in human history, acting as catalysts for scientific discovery, shaping cultural narratives, and even changing the result of battle. They serve as a reminder of the interconnectedness of the universe and human civilization, eliciting both terror and amazement. As we continue to monitor and understand these cosmic phenomena, they will remain an important part of our shared human experience.

Myths and Legends about Eclipses

Eclipses have long been a source of mystery and awe, inspiring a wide range of myths and stories throughout countries and periods. These astronomical events, in which the Sun or Moon appears to vanish, have frequently been perceived as potent omens, resulting in myths that attempted to elucidate the phenomenon in terms comprehensible to the people of the time.

The Dragon that devours the Sun

In many Asian civilizations, a frequent theme is a heavenly dragon or beast devouring the Sun during an eclipse. In ancient China, it was thought that a dragon swallowed the Sun, so people banged drums and made loud noises to scare the dragon away and free the Sun from its jaws.

Sun and Moon in Conflict

According to Inuit tradition, the Sun goddess Malina walked away following a quarrel with the Moon deity Anningan, causing the Sun to darken during an eclipse. This myth depicts the Inuit's view of the eclipse as the outcome of a fight between heavenly creatures.

The Hindu Demon Rahu

According to Hindu legend, the demon Rahu, who was decapitated by the gods for drinking the nectar of immortality, chases the Sun and Moon out of vengeance. During an eclipse, Rahu is said to seize and consume one of them, only to resurface after passing through his severed neck.

Pomo Bear and the Sun

The Pomo, an indigenous community from the northwest United States, tells a story about a bear who fights the Sun and bites it, causing a solar eclipse. This story explains how eclipses occur and their cyclical nature, as the bear finally bites the Moon, resulting in a lunar eclipse.

Vikings and Wolves

Norse mythology provides a distinct perspective, with the wolves Sköll and Hati chasing the Sun and Moon, respectively. During an eclipse, it is thought that one of the wolves has caught and is attempting to consume the heavenly body, which adds to the Norse interpretation of the event.

The Vietnamese Frog and Sun

In Vietnam, there is a mythology that a huge frog swallows the Sun, causing solar eclipses. This tale represents an attempt to explain the Sun's rapid departure throughout the day.

The choctaw Black Squirrel

The Choctaw people of North America relate the story of a black squirrel attempting to devour the Sun, resulting in a solar eclipse. This fable exemplifies how natural events were frequently explained by the acts of animals in Native American civilizations.

Greek Omens

The ancient Greeks saw a solar eclipse as a portent of angry gods and the start of tragedies and devastation. This theory emphasizes how eclipses were frequently viewed as signs of doom in numerous societies.

Modern interpretations

Eclipses might still cause some individuals to feel uneasy or superstitious. However, they continue to fascinate and inspire writers and artists. The myths and tales surrounding eclipses have changed over time, yet the sense of awe they evoke stays constant.

The myths and tales about eclipses are as diverse as the societies that inspired them. From dragons and devils to bears and squirrels, these stories illustrate humanity's efforts to comprehend and explain the workings of the universe. They also serve as a reminder of our shared urge to understand the world around us, demonstrating the power of narrative in our search for knowledge.

Chapter 3: Preparing for the Eclipse

Choosing the Best Location

Choosing the greatest place to watch a solar eclipse is an important component of the planning process. The experience might differ greatly depending on where you are located.

Understanding the path of totality

The path of totality is the narrow strip where the Moon covers the Sun, and it is the only location where you can see the entire spectacle of a total solar eclipse. This trail may be as wide as 200 kilometers and covers immense areas. To see the eclipse in its entirety, you must be on this path.

Geographic considerations

When deciding on a site, consider the following geographical factors:

Accessibility: Select a place that is accessible via several forms of transportation. Remote sites may provide a better view, but they might be difficult to access, especially with the rush of eclipse chasers.

Altitude: Higher altitudes may give a better viewing experience because there is less atmosphere to block the view.

Proximity to the Center Line: The closer you are to the centerline of the path of totality, the longer the complete eclipse will last.

Weather and Climate

Weather is an important aspect of eclipse viewing. Clear skies are required for an uninterrupted view. To select a site with the highest chances of clear sky, look into historical weather trends and long-term projections. Some areas within the line of totality are more likely to see favorable weather than others.

Duration of totality

The duration of totality varies according to the path. Locations in the center line will have the greatest length, while those on the periphery will have a shorter totality phase. Choose a location that allows the Sun to be shaded for the longest amount of time.

Local Infrastructure

Consider the local infrastructure in your selected area. Areas with greater facilities may offer a more pleasant experience. This covers hotel, meals, transportation, and emergency services.

Crowd Management

Popular viewing areas might grow crowded, affecting your experience. Look for sites that are projected to be less congested or that have stronger crowd control measures in place.

Community Events

Many villages along the road of wholeness hold unique events, festivals, and educational initiatives. Participating in these events can enhance your eclipse-viewing experience.

Safety precautions

Safety should be the first focus. Ensure that you have the appropriate solar viewing glasses that fulfill international safety requirements. Be mindful of your surroundings, especially if you're in a rural location.

Practice Runs

If feasible, pay a visit to your preferred place beforehand. This might help you get to know the region and choose the finest viewing location.

Backup Plans

Always keep a backup plan. If the weather becomes bad or other unexpected circumstances happen, be prepared to relocate to another place.

Choosing the greatest spot to see a solar eclipse requires careful consideration of several criteria, including geography, weather, length of totality, local infrastructure, and safety. A remarkable and awe-inspiring eclipse experience may be achieved by planning ahead of time and making good choices.

Prepare for the upcoming complete solar eclipse on April 8, 2024, to fully appreciate this unique celestial spectacle.

Equipment Checklist.

Preparing for a solar eclipse is an exciting activity, and having the proper equipment is critical to a safe and pleasurable viewing experience. Here's a comprehensive checklist to ensure you're ready for the upcoming solar event.

1. Eclipse viewing glasses

The most critical item on your checklist should be eclipse glasses that adhere to the international safety standard ISO 12312-2:2015. These glasses are intended to shield your eyes from dangerous solar radiation and lower the Sun's brightness to a safe and pleasant level. Make sure your eclipse glasses are not damaged or scratched; even a little rip might diminish their efficiency.

2. Solar Filters for Telescopes and Binoculars

If you intend to use binoculars or a telescope, you will require sun filters that fit over the front of your optics. These filters limit the bright sunlight before it enters the telescope, safeguarding both your eyes and the equipment. When viewing an eclipse, never use ordinary sunglasses or unfiltered cameras, telescopes, or binoculars.

3: Pinhole Projector

A pinhole projector is a safe and simple way to see objects indirectly. You may make one out of cardboard with a small hole that projects the Sun's image onto a flat surface. This allows you to follow the eclipse's progress without gazing straight at the sun.

4. Tripod for Stability

A tripod is necessary if you're using a camera, binoculars, or a telescope. It offers stability and allows you to retain the same viewing angle during the eclipse.

5. Camera Equipment

If you intend to shoot the eclipse, make sure you have a manual camera with a telephoto lens and a certified solar filter for the camera lens. To keep your camera stable, you'll also need a tripod.

6. Extra lenses and batteries
Bring several lenses with varying focal lengths and extra batteries to guarantee you don't miss any of the eclipses due to power outages.

7. Remote Shutter Release.
A remote shutter release permits you to snap shots without touching the camera, which decreases the probability of camera shaking and fuzzy images.

8. Chair, Blanket
Comfort is essential during the eclipse. Bring a chair and a blanket so you can relax and enjoy the event.

9) Food and Water
Eclipse watching is frequently a waiting game. Pack extra food and drink to stay hydrated and energetic.

10) Notebook and Pen

Keep a journal and pen on hand to record times, observations, and any unusual experiences throughout the eclipse.

11. Portable Power Bank

A portable power bank can keep your electronics charged during the event.

12. First Aid Kit.

Always keep a basic first aid kit on hand in case of minor accidents or health concerns.

13. Flashlight.

As totality approaches, it will become darker. A flashlight can assist you in securely exploring the viewing area.

14. Weather Appropriate Clothing

Check the weather forecast and dress appropriately. Temperatures may dip during totality, therefore layers are suggested.

15. Map and Information

Have maps and information on the eclipse's course and timing so you can be in the correct spot at the right moment.

16. Eclipse Timing Application

Download an eclipse timing app to keep track of the eclipse's stages and length.

17. Back-up Plan

Prepare a backup site in case the weather becomes adverse at your prime viewing area.

By completing this checklist, you'll be well-equipped to witness the solar eclipse safely and comfortably. Remember that preparation and safety are crucial for a tremendous eclipse viewing experience.

Plan Your Eclipse Trip

Planning a journey to see a solar eclipse is an adventure that combines the excitement of travel with the wonder of astronomy. As we approach the complete solar eclipse on April 8, 2024, it is critical to methodically arrange for a memorable experience. Here's a thorough guide on preparing for your eclipse travel.

Understanding the path of totality

The path of totality is the best viewing position since the eclipse is total and the Sun is entirely veiled by the moon. This route is often a thin band that spans the Earth's surface. The 2024 eclipse will span North America, providing a unique opportunity for millions to see this uncommon occurrence.

Selecting Your Destination

It is critical to choose the appropriate destination along the road to wholeness. Consider accessibility, past weather trends, local infrastructure, and probable crowds. Look for sites with clear skies and the logistical capabilities to accommodate a large number of guests.

Booking Accommodations

Accommodations in popular areas tend to fill up fast, so book your stay in advance. Explore a variety of accommodations, from hotels to campers, and consider staying in surrounding towns if the key locations are filled. Make sure your hotel has flexible cancellation procedures in case your plans alter due to the weather or other unforeseen situations.

Transportation

Plan your travel to and within the destination. If you're flying, plan early to avoid exorbitant fares and limited availability. If you are driving, make sure your car is in good shape and you are familiar with the route. Consider traffic patterns on the day of the eclipse and arrange alternative routes.

Eclipse timing

Understand the timing of the eclipse stages. Knowing when the partial phase begins, when totality occurs, and how long it lasts at your location can allow you to better organize your day. To obtain precise time information, use trusted sources or applications.

Safety Gear

Eclipse glasses are a must-have for safeguarding your eyes while watching the eclipse. Ensure that they fulfill the international safety standard ISO 12312-2:2015. If you intend to utilize cameras or telescopes, be sure they are equipped with appropriate sun filters.

Practice observing

If possible, practice watching the Sun with your eclipse glasses and any equipment you intend to bring. Familiarity with your gear will improve your watching experience and allow you to capture the event securely and efficiently.

Weather Conditions

The weather is unpredictable, so make a backup plan. Determine other viewing places with good weather forecasts that you may visit if required. Get real-time weather updates as the eclipse approaches.

Culture and Community Events

Special events and activities will take place at several points along the entire trail. Participate in local cultural activities, discussions, and gatherings to enhance your eclipse experience.

Physical comfort

Prepare for the physical aspects of your vacation. Bring appropriate attire for the weather, sunscreen, drinks, snacks, and a first-aid kit. Consider the amount of time you'll be outside and plan accordingly.

Photography & Equipment

If you want to shoot the eclipse, pack the necessary equipment, which includes a camera with a solar filter, a tripod, and additional batteries. To prepare for the occasion, practice snapping images under various lighting circumstances.

Leave no Trace

Be conscious of the environment. Follow the Leave No Trace guidelines to guarantee that the natural sites you visit are preserved for future tourists and wildlife.

Post-Eclipse Activity

After the eclipse, spend some time exploring the region. Many places have distinctive attractions, natural beauty, and historical monuments. Extend your journey to see what else the region has to offer besides the eclipse.

Many elements must be carefully considered while planning an eclipse vacation, from selecting the best viewing position to assuring your safety and comfort. Following these recommendations will prepare you to watch one of nature's most stunning shows. With proper planning, you may witness the 2024 complete solar eclipse from a front-row seat.

Chapter 4: Safety First

How to safely see a solar eclipse

Seeing a solar eclipse is an exciting event, but you must do so properly to safeguard your eyes from potential injury.

Understanding the Solar Eclipse

A solar eclipse happens when the Moon passes between the Earth and the Sun, briefly covering its light. Solar eclipses are classified into three types: total, annular, and partial. During a partial eclipse, the Moon only covers a portion of the Sun. An annular eclipse occurs when the Moon obscures the Sun's center, leaving a ring of light visible. A total eclipse fully obscures the Sun, displaying its corona.

The Importance Of Eye Safety

The Sun generates powerful visible light as well as invisible ultraviolet (UV) and infrared (IR) radiation. Viewing the uneclipsed or partially eclipsed Sun without suitable protection can result in "eclipse blindness" or retinal burns, which can lead to lifelong eye injury or blindness.

Eclipse glasses

The best approach to seeing a solar eclipse is using eclipse glasses that fulfill the international safety standard ISO 12312-2: 2015. These glasses have special-purpose solar filters that block dangerous UV and IR radiation while reducing the Sun's brightness to a safe, pleasant level. Before using your glasses, check for scratches, punctures, rips, or damage. If they are broken, they should be disposed of. Never use conventional sunglasses, no matter how black they are, since they transmit millions of times more sunlight.

Pinhole projectors

A pinhole projector is suitable for indirect viewing. This DIY approach projects the Sun's image onto a surface via a pinhole, allowing you to observe the eclipse without staring directly at it. To make one, simply punch a small hole in a card and place it between the Sun and a screen (such as a sheet of paper), and an image of the Sun will come up on the screen.

Telescope and binoculars

If you want to use a telescope or binoculars, be sure they have an appropriate solar filter on the front lens. These filters prevent direct sunlight before it enters the instrument. Never stare through a piece of unfiltered optical equipment, since this might concentrate sun radiation and inflict serious harm.

Photographing the eclipse

If you want to picture the eclipse, you'll need a camera with a solar filter on its lens. Use a tripod to keep your camera stable, and a remote shutter release to reduce shaking. Remember to remove the filter only for a brief period of totality.

During the eclipse

During the partial stages of the eclipse, constantly stare at the Sun via your eclipse glasses. If you're in the path of totality, you can safely remove your glasses after the Moon has fully covered the Sun. Put on your spectacles as soon as the sun reappears.

Children and Eclipse Viewing

Ensure that youngsters understand the necessity of using eclipse glasses and that they do so appropriately at all times.

Pets and eclipses

Pets seldom gaze directly at the Sun, thus they do not normally require protection. If you are concerned about eye harm, keep children indoors during the eclipse.

Protecting your eyes

Protecting your eyes is critical not just during special occurrences like solar eclipses, but also in everyday life. Our eyes are delicate organs that require attention to ensure proper vision and general health.

Understanding Eye Hazards

Various environmental elements might harm our eyes. Ultraviolet (UV) radiation from the sun, blue light from screens, chemicals, dust particles, and physical stress from sports or work-related activities can all cause eye injury if not handled properly.

UV Radiation Protection

Exposure to UV radiation can cause cataracts, macular degeneration, and photokeratitis (corneal sunburn). To safeguard your eyes from ultraviolet rays:

Wear sunglasses that completely filter UVA and UVB radiation.

Wear a wide-brimmed hat to protect your eyes from direct sunshine.

Avoid gazing directly at the sun, especially during a solar eclipse, unless you wear ISO-certified eclipse glasses.

Bluelight and Digital Eye Strain

With the increased usage of digital gadgets, our exposure to blue light has increased dramatically. Prolonged exposure can cause digital eye strain, which is characterized by dryness, discomfort, and weariness. To lessen the effects:

Apply blue light filters to screens.

Follow the 20-20-20 rule, which states that every 20 minutes, stare at anything 20 feet away for at least 20 seconds.

Use suitable lighting to prevent glare on displays.

Chemical and Particle Protection

Chemicals and tiny particles can cause severe eye damage. If you deal with dangerous products or in areas with flying debris:

Wear safety goggles or face shields that fulfill safety requirements.

Understand the emergency protocols for eye cleaning in the event of chemical exposure.

Ensure that eyewash stations are accessible and well-maintained.

Sports and Physical Activities

Sports-related eye injuries are prevalent but may be prevented. If you're playing basketball, swimming, or cycling:

Wear protective eyewear with polycarbonate lenses.

Select eyeglasses suited for the sport.

Ensure that youngsters engaging in sports have eye protection.

Regular Eye Examination

Regular eye examinations are important for maintaining eye health. They can identify early indicators of eye diseases that do not have evident symptoms. Comprehensive eye exams are advised every one to two years, based on age and health.

See an eye care expert right away if you have abrupt vision changes, discomfort, or damage.

Healthy Lifestyle Options

Your entire wellness affects your eye health. To safeguard your vision, eat a balanced diet high in vitamins and antioxidants.

Stay hydrated to keep your eyes moist.

Quit smoking, as it raises the risk of eye illness.

First Aid for Eye Injury

Basic first aid for eye injuries can help avoid additional harm. If you or someone else has an eye injury, avoid rubbing or applying pressure to the affected eye.

If a foreign object is present, do not try to remove it yourself.

Seek expert medical assistance immediately.

Eye Protection at Home

Even at home, it's necessary to consider eye safety:

Keep any home chemicals and sharp items out of reach of youngsters.

Be cautious while cooking with oil or grease to avoid splatters.

Exercise caution while opening champagne bottles or using instruments that may launch things.

Workplace Eye Safety

When working in an atmosphere with probable eye hazards, it's important to comprehend the dangers and wear convenient eye protection.

Make sure your protective eyewear fits properly and is comfortable.

Participate in any eye safety training offered by your workplace.

Eye safety should be a top priority in all parts of life. Understanding the dangers and taking the required measures can help you safeguard your vision and keep it healthy for years to come. Remember that when it comes to your eyes, safety first is more than just a slogan; it is a must for maintaining one of your most precious senses.

Common Misconceptions and Dangers

Misconceptions about solar eclipses can lead to wasted opportunities as well as major health dangers. Understanding these frequent myths and risks is critical for anybody planning to see an eclipse. Here's an in-depth look at some of the most common fallacies and the actual hazards they provide.

Every Solar Eclipse is the Same

Many people assume that once they have witnessed one solar eclipse, they have seen them all. This is far from true. Solar eclipses come in four types: partial, annular, total, and hybrid, and each provides a unique experience. A total solar eclipse, in which the Moon covers the Sun, is a rare occurrence that may be deeply moving owing to the sudden night-like darkness and view of the Sun's corona.

A 99% Partial Eclipse Is Almost Equal to Totality

There's a huge difference between a 99% partial eclipse and totality. Even a sliver of the Sun's dazzling disk can be too bright for the eyes, and the distinctive phenomena associated with totality, such as the development of the corona and "Baily's beads," are not visible during a partial eclipse.

It's Never Safe to View a Solar Eclipse

It is only safe to stare at the Sun with the naked eye during the brief period of totality in a complete solar eclipse. To avoid significant eye injury, use good eye protection at all times, even during partial and annular eclipses.

Solar Eclipses Harm Pregnant Women and Unborn Children

There is a longstanding belief that solar eclipses can damage pregnant women and their unborn babies. However, no scientific data supports this assumption. The notion is most likely rooted in old assumptions and superstitions that correlate eclipses with terrible omens.

If you are not in the direction of totality, you do not need eye safety

Even if you're not in the line of totality, staring at the sun without adequate protection might cause irreversible eye damage. Always use ISO-certified eclipse glasses or other solar viewing techniques when watching an eclipse.

Real Risks of Observing a Solar Eclipse

The true risk of a solar eclipse is to your eyes. UV radiation from the Sun can induce "eclipse blindness" or retinal burns, also known as solar retinopathy. Because the retina lacks pain receptors, this damage can occur even when there is no discomfort. The consequences might be transient or permanent, resulting in partial or complete blindness.

How to safely see a solar eclipse

To safely view a solar eclipse, follow these guidelines:

Wear Proper Eye Protection: Always wear eclipse glasses that comply with the ISO 12312-2 international safety standard. Regular sunglasses, even extremely black ones, are not suitable for staring at the sun.

Supervise the Children: Make certain that youngsters use eclipse glasses appropriately and understand the significance of never staring straight at the Sun without them.

Use Indirect Viewing Methods: If you don't have eclipse glasses, you can observe the eclipse with a pinhole projector.

Do not use unfiltered optics: Never gaze at the Sun with unfiltered cameras, telescopes, binoculars, or other optical instruments.

Be Cautious with Photography: If you're shooting the eclipse, make sure your camera has a sun filter to protect both your eyes and the sensor.

Solar eclipses are wonderful spectacles that should be appreciated responsibly. By debunking popular myths and knowing the true risks, spectators may fully experience the majesty of an eclipse without jeopardizing their health. Remember to prioritize safety during solar eclipses.

Chapter 5: Photography and the Eclipse

Photographing the Eclipse - Tips and Techniques

Photographing the eclipse is an intriguing challenge for both amateur and professional photographers. A solar eclipse provides a unique opportunity to picture a rare celestial occurrence, but it must be carefully planned and executed. Here's a complete guide on shooting solar eclipses.

Understanding Eclipse

Before you begin organizing your photographs, you must first grasp the eclipse's stages and their time. The eclipse will go through partial stages, culminating in totality, before reversing course. Each era has unique photography opportunities and challenges.

Equipment Checklist - Camera: A DSLR or mirrorless camera with manual mode is recommended.

Lenses: A telephoto lens (at least 200mm) is suggested for taking detailed photographs of the eclipse.

Tripod: Use a strong tripod to support your camera and reduce camera shake during extended exposures.

sun Filter: A sun filter is required to protect your camera sensor and your eyes when capturing the partial stages of the eclipse.

Extra Batteries and Memory Cards: Be sure you have enough power and storage for the event.

Remote Shutter Release: This reduces vibrations when shooting images.

Camera Settings

Aperture: Choose a mid-range aperture (f/8 to f/16) for a crisp photograph of the sun.

Shutter Speed: Set a fast shutter speed (1/1000 second) and adjust as needed. During totality, greater shutter speeds may be required to catch the features of the corona.

ISO: Set the ISO low to reduce noise. Increase it only if necessary during totality.

Focus: Before the eclipse, set your lens to manual focus and pre-focus on the sun.

White Balance: Change your white balance to daylight to keep a constant color temperature.

Photographing Partial Phases

During the partial phases, you must apply a sun filter to protect your camera and eyes. Photograph the sun at regular intervals to record the moon's journey across it. Remember to adjust your exposure as the sunshine fades.

Capturing the Totality

Only at totality is it safe to shoot the eclipse without using a sun filter. Remove the filter when the sun is entirely obscured and catch the breathtaking corona. Experiment with various exposures to capture the changing intensity of the corona's brightness.

Safety first

Never stare directly at the sun via your camera's viewfinder without a solar filter. Use the live view mode instead. Always put the sun filter back on as soon as the totality is over.

Practice Makes Perfect

Practice photographing the sun in the days preceding the eclipse (with a solar filter). Familiarize yourself with your equipment and the settings you will be using.

Composition and Framing

Consider the composition of your eclipse photographs. Will you incorporate foreground items for context, or will you only focus on the sun? Plan your framing, and consider choosing a longer lens for a closer look or a broader lens for a landscape image.

Post-Processing

After the eclipse, you'll most likely have a collection of photographs depicting the many stages. Post-processing may help you improve contrast, highlight details, and even produce a composite image that depicts the eclipse's evolution.

Photographing a solar eclipse is a fulfilling experience that needs planning and attention to detail. Following these ideas and approaches will prepare you to take spectacular photographs of this cosmic spectacle. Remember to emphasize safety, rehearse beforehand, and enjoy the experience of recording a moment in cosmic history.

Required Equipment for Eclipse Photography

Photographing a solar eclipse is an incredible opportunity for both amateur and professional photographers. To catch this cosmic event, you'll need the proper equipment and understanding. Here's a complete guide on the equipment required for eclipse photography.

Camera body

A DSLR or mirrorless camera with manual settings is needed. For the best results, you must be able to manually modify focus and exposure settings, as well as shoot in RAW format.

Lenses

A telephoto lens is essential for taking detailed pictures of the eclipse. A focal length of at least 200mm is ideal, while longer lenses can give more detail. Consider using a lens with a fast aperture to collect more light during totality.

Tripod

A robust tripod is required to support your camera, particularly when using long telephoto lenses. It will also assist you in maintaining composition and focus during the eclipse's many stages.

Solar filters

Solar filters are essential to protect your camera's sensor and your eyes when photographing the sun during a partial eclipse. These filters must be particularly made for sun photography and should be applied to the front element of your lens.

Additional Batteries and Memory Cards

Eclipse photography may be battery-intensive, especially if you want to utilize live view or record a time-lapse. Bring additional fully charged batteries and enough memory card space to make sure you don't miss anything throughout the event.

Remote shutter release

A remote shutter release allows you to snap shots without directly touching the camera, which reduces camera shaking and ensures clear images.

Viewing Glasses

Eclipse glasses are required to safely observe the eclipse phases with your eyes. Ensure that they fulfill the ISO 12312-2 international safety standard.

Intervalometer

If you want to make a time-lapse of the eclipse, an intervalometer will come in handy. It will allow you to shoot a series of images at predetermined intervals without requiring manual intervention.

Lenses Hood

A lens hood can assist in reducing lens flare and protect the front element of your lens from scratches or damage.

Neutral density filters

Neutral-density filters can be handy for managing exposure, especially if you intend to photograph the scenery alongside the eclipse.

Flashlight

A modest flashlight or headlamp can be quite useful while setting up your equipment in the dark or low-light situations during totality.

Chairs and Comfort Items

Bring a comfortable chair, food, and drink since you will be waiting for long periods. Being comfy might help you stay focused on your photography.

Weather protection

Prepare for any weather by packing lens cloths, rain coverings for your camera, and proper clothes.

Preparation and Practice

Prepare for the eclipse by familiarizing yourself with your equipment. Practice setting up and disassembling your equipment, adjusting settings fast, and replacing batteries and memory cards.

To photograph the eclipse, you must be prepared and have the proper equipment. You'll be prepared to shoot this breathtaking occurrence if you have a powerful camera, adequate lenses, sun filters, and other necessary equipment. Prepare for the eclipse by practicing with your equipment ahead of time.

Chapter 6: The Eclipse Experience

What to expect during the eclipse

Witnessing a solar eclipse is an unforgettable experience that mixes the predictable orbits of celestial bodies with the unexpected emotions of the human soul. Here's what to anticipate during the eclipse, from scientific occurrences to potential emotional responses.

The Approach

As the moon begins its passage over the sun, the day will gradually fade, throwing an eerie twilight over the countryside. The temperature will decrease, and animals may become confused, believing night has arrived early. This gradual alteration in the surroundings is the first indication that something remarkable is occurring.

First contact

The eclipse begins with the first contact when the moon makes its initial pass over the sun. Through safe viewing methods, you may witness the sun gradually take on a crescent form as the moon covers it. This period might take more than an hour, depending on your location.

Crescent Sun

As the moon covers more of the sun, the crescent form becomes more visible. Shadows will become sharper, and the quality of light will shift. The world will appear to be viewed through a filter, with colors becoming more saturated and contrasts increasing.

Baily's Beads with Diamond Ring

Just before totality, the uneven lunar topography permits beads of sunshine to show through in certain areas but not in others. These are called Baily's Beads. The beautiful "diamond ring" appearance is created as the beads vanish and the last dazzling patch of sunlight shines through.

Totality

Totality marks the culmination of the eclipse. For a short few minutes, the moon fully obscures the sun. The corona, the sun's outer atmosphere, becomes visible, forming a white halo around the darkening moon. Stars and planets may become visible, and the horizon shines with sunset hues from 360 degrees around.

Shadow bands

Another thing to look for is elusive shadow bands wavy lines of alternating light and dark that may be seen moving on plain-colored surfaces shortly before and after totality. They result from the sun's light being twisted and focussed by the Earth's atmosphere.

Emotional Experience

Many people have emotional reactions to eclipses. Some people express a strong sense of connectedness to the universe, while others are touched by the pure beauty of the occurrence. The abrupt blackness of totality might feel otherworldly, while the reappearance of the sun can feel like a rebirth.

Safety

Throughout the eclipse, it is critical to protect your eyes. Only during the brief period of totality is it safe to stare at the sun without protection. To avoid eye injury, you must always use eclipse glasses or utilize alternative sun-viewing methods.

Photography

Preparation is required for individuals who want to capture the occasion through photography. You'll need a solar filter for your camera and an understanding of how to modify settings for changing light levels. Practice before making sure you're prepared.

The Return of Light

As the moon travels away from the sun, regular daylight eventually returns. The temperature rises, animals return to their normal behavior, and the atmosphere feels familiar once more. However, the eclipse leaves a lasting impression in the form of recollections and images.

Personal Accounts for Totality

Witnessing totality during a solar eclipse is a memorable experience for many people. It's the transition from day to night when temperatures drop and stars may emerge in the sky during midday. Here's a thorough examination of human experiences and the powerful influence of wholeness.

The anticipation

The buildup to totality is fraught with expectation. Observers frequently report increased excitement as the moon begins to obscure the sun. The globe gradually dims, and temperatures begin to plummet. Animals and birds may go silent or act as if darkness has fallen, contributing to the bizarre mood.

A Changing Environment

As the eclipse proceeds, the quality of light varies. Shadows grow crisper, and the landscape takes on a silvery tone. Observers frequently note a weird stillness in the air, a stop in time that is both soothing and unsettling.

Emotional Impact

The beginning of completeness might elicit a variety of feelings. Some people express a profound sensation of wonder or a strong connection to the universe. Others express sensations of exhilaration, terror, or even apocalyptic visions. The blackness of totality, along with the beautiful corona of the sun, provides a vision that may bring people to tears.

The Diamond Ring & Baily's Beads

Just before and after totality, the diamond ring effect—a spectacular flash of light as the sun peeks around the moon—and Baily's Beads—points of light streaming through the moon's valleys—provide a stunning visual spectacle. These occurrences are frequently mentioned in personal reports as among the most memorable features of the eclipse.

Shared Experience

Totality is sometimes referred to as a community event. People form communities to view the eclipse, exchanging equipment, expertise, and feelings. The communal experience of totality can foster a sense of community among strangers who have seen a rare natural phenomenon.

The moment of totality

The totality is often fleeting, lasting barely a few minutes. However, in those minutes, the world is altered. The sun's corona becomes apparent, forming a halo of light around the darkened moon. Observers frequently use terms like "otherworldly" and "ethereal" to describe their appearance.

The Return of the Light
As the moon travels away, the sun reappears, bringing daylight back. Observers typically express relief or a sense of rejuvenation as the sun returns. The end of totality can also cause regret that the moment has passed, as well as a wish to relive it.

The Long-Lasting Effects
Many people who saw totality discuss the event's long-term ramifications. It can alter one's perception, making the cosmos appear both large and close. Some find it inspiring to study astronomy, while others see it as a reminder of nature's beauty and majesty.

Personal experiences of totality during a solar eclipse describe a profound and transformational experience. It's a moment where the predictable physics of heavenly bodies meet the unexpected human response to the sublime. For those who have observed it, totality is more than simply a visual spectacle; it is an emotional journey and a deeply personal event that lasts long after the eclipse.

These testimonies demonstrate the unique and personal character of seeing a complete solar eclipse. Each individual's experience is unique, yet all stories of wholeness share a common thread of awe and connection. It's an extraordinary occurrence that will leave an unforgettable impact on the hearts and minds of those who witness it.

The Emotional Effects of Eclipses

Eclipses have had a strong and diverse emotional influence on mankind for millennia. Eclipses, particularly total solar eclipses, are among the most spectacular astronomical occurrences observable from Earth, eliciting a broad range of emotions, including awe and amazement, terror, and contemplation. This article investigates the emotional reactions evoked by eclipses, the psychological impacts they can have, and how they alter our perceptions of the world and ourselves.

Awe and Wonder

The most typical emotional response to an eclipse is wonder. Awe is a complex emotion that includes sentiments of surprise, veneration, and even terror in the presence of something immense and beyond human comprehension. Eclipses, with their majesty and rarity, naturally elicit this emotion.

Observers frequently report feeling overwhelmed by the magnitude of the event, as the normal order of the day is suddenly disrupted and the sun is eclipsed by the moon.

Connection with the Cosmos

Eclipses may also provide a sense of connectedness with the universe. As the day goes to night and back again in only a few hours, we are reminded of our place in the cosmos and the bigger forces at work. This cosmic viewpoint may foster sentiments of togetherness with nature and mankind, as people all across the world see the eclipse.

Reflection and introspection

The uncommon and spectacular character of an eclipse can cause people to ponder on their own lives and the world around them. The momentary darkness of a total solar eclipse can serve as a metaphor for life's uncertainties and difficulties. It might prompt reflection on one's personal objectives, relationships, and role in the world.

Fear and Superstitions

Eclipses have always inspired dread and superstition. Many ancient societies saw them as omens or messages from the gods, frequently predicting imminent calamity. While most people's worries about astronomy have been alleviated by contemporary knowledge, the abrupt blackness of an eclipse can nevertheless elicit a primeval sensation of dread or worry in others.

Transformation & Renewal

For many people, an eclipse represents transition and rejuvenation. The eclipse's cycle, from initial contact to full sunshine, may signify both ends and new beginnings. This symbolism is frequently used by people seeking transformation or improvement in their lives.

The social impact of eclipses

Eclipses may have a large societal influence as well. They gather people, usually in huge groups, to observe the event. This common experience may generate a sense of belonging and enthusiasm. The wonder felt by an eclipse can lead to enhanced socialization, humility, and a sense of connectedness to others.

Psychological Effects of Awe

The sense of awe during an eclipse can provide tangible psychological benefits. Awe may boost our curiosity, openness, and creativity. It can also make us more kind and empathetic since it typically gives us a sense of belonging to something bigger than ourselves.

Eclipses of the Modern World

Eclipses continue to amaze and inspire people today. They provide opportunities for scientific inquiry, creative expression, and public celebration. The emotional effect of an eclipse may be increased by the media, as photographs and tales are disseminated internationally, allowing people who are unable to view the event in person to engage in the collective experience.

Eclipses have an equally broad and strong emotional influence. These cosmic phenomena may provoke a broad range of emotions, including awe and amazement, terror, and reflection. They remind us of our place in the universe, encourage us to reflect on our lives, and bring us together via shared experiences. Eclipses continue to inspire and impact us emotionally as we observe and study them.

www.ingramcontent.com/pod-product-compliance
Lightning Source LLC
Chambersburg PA
CBHW070348230526
45471CB00006B/2468